CREATURES
BY DESIGN

DISCOVERING CHRIST'S INCREDIBLE
ENGINEERING IN ANIMALS

BRIAN THOMAS · FRANK SHERWIN
RANDY J. GULIUZZA · JAMES J. S. JOHNSON

INSTITUTE FOR
CREATION
RESEARCH

Dallas, Texas
ICR.org

CREATURES BY DESIGN
Discovering Christ's Incredible Engineering in Animals

by Brian Thomas, Ph.D., Frank Sherwin, D.Sc. (Hon.),
Randy J. Guliuzza, P.E., M.D., and James J. S. Johnson, J.D., Th.D.

First printing: July 2021

All Scripture quotations are from the New King James Version.

ISBN: 978-1-946246-61-5

Please visit our website for other books and resources: ICR.org

Printed in the United States of America.

TABLE OF CONTENTS

1

MONARCH BUTTERFLY ANTENNA, A TINY HIGH-TECH TOOLKIT

A 3,000-mile road trip would prove daunting to many humans, but monarch butterflies can migrate that distance every fall from Canada to a specific grove of fir trees in Mexico. The next generation of monarchs then has the innate ability to make the return trip in the spring.

Scientists are investigating the tools these tiny flying creatures use to achieve this feat. One leading monarch researcher discovered an important reason why the butterflies' antennae are vital for successful navigation. Neurobiologist Steven Reppert and his team wanted to find the specific mechanism in the antenna that enables the butterfly to migrate accurately.

In a report on their 2009 study published in *Science*, Dr. Reppert stated, "We've known that the insect antenna is a remarkable organ, responsible for sensing not only olfactory cues but wind direction and even sound vibration."[1] After observing how removing or painting over the antennae affected navigation, they found yet another sensor tucked in the tiny tendril—one that detects the angle of incoming sunlight.

In 2005, Dr. Reppert discovered that monarch eyes detect a UV portion of sunlight that coordinates with a circadian clock in the brain.[2] Four years later he was surprised to find a clock, or "time correction factor," housed inside the antenna itself.[1] The internal clock constantly calibrates position using the angle of sunlight, converting that information into cardinal directions. That way the butterflies can fly south or north unerringly.

These butterflies, therefore, have a well-organized network of data coordination. Both eyes and antennae detect light information, and programmed logic centers integrate those data with circadian clocks in the brain and antennae. After considering the chemical, wind, and sound detectors within the antennae—along with the geographic information that is somehow transferred to the second generation to enable them to travel to a location they've never been—one can see that such specifications could never be generated randomly.

The butterfly's navigation system resembles the sophisticated autopilot systems on aircraft. These systems sense and track factors like wind speed, altitude, and destination coordinates, then use programmed logic centers to keep the plane flying straight. The human design we clearly see in autopiloting systems leaves no doubt about divine design in butterflies.

For it to function properly, an engineer would need to specifically engineer the butterfly to work in conjunction with the complex systems present in its eyes and brain to steer it to its programmed destination. And what better candidate for the identity of this engineer than the Lord Jesus Christ? Creation explains the clearly seen design in monarch navigation systems.[3]

References

1. Migrating Monarch Butterflies "Nose" Their Way to Mexico. University of Massachusetts Medical School press release, September 25, 2009, reporting research published in Merlin, C., R. J. Gegear, and S. M. Reppert. 2009. Antennal Circadian Clocks Coordinate Sun Compass Orientation in Migratory Monarch Butterflies. *Science*. 325 (5948): 1700-1704.

2. How butterflies fly thousands of miles without getting lost revealed by researchers. The Hebrew University of Jerusalem press release via Eurekalert! August 1, 2005, reporting research published in Sauman, I. et al. 2005. Connecting the Navigational Clock to Sun Compass Input in Monarch Butterfly Brain. *Neuron*. 46 (3): 457-467.

3. Genesis 1:31.

2

THE PANGOLIN, A MAMMAL WITH SCALES

Pangolins (scaly anteaters) are uniquely engineered armored mammals found in southeast Asia and Africa. The pangolin's face is like an armadillo's. Its head is tubular-shaped, and the fused jaws are without teeth.

Pangolins walk on their hind legs, but when threatened they roll themselves into a tight armored ball. Their robust layers of keratin scales make them look a like a cross between an artichoke and a pineapple. These formidable scales make up about 20% of the pangolin's body weight. Pangolins are designed to slurp up insects at a voracious rate—about 60 million ants per year.

Jesus Christ engineered the giant pangolin (genus *Manis*) as a fearsome eating machine with a long, sticky, narrow tongue that's actually longer than its body—about three feet. Like in the woodpecker, the tongue must be specially stored when it's not being used. The pangolin retracts the tongue into a sheath in its chest cavity. Within this casing are large glands that lubricate the yard-long tongue with special saliva so the creature can chow down on termites and ants. The muscles operating the tongue are attached to the elongated xiphoid process of the breastbone.

The stomach is engineered with keratinous spines and contains gravel to grind the insects, which is similar to the muscular gizzard found in chickens and turkeys. Digestion occurs in the relatively long small intestine. While it forages in termite mounds it can close its nostrils and ears, keeping the angry insects out.

As is always the case with animal origins, pangolin "ancestry has been surprisingly disputed." Evolutionary paleontologist Michael Benton also states these fossils "are surprisingly modern-looking."[1] This might surprise the Darwinist but not the creation scientist who maintains pangolins have always been pangolins. Indeed, paleontologists found 100% pangolin fossils from the early Eocene Messel in Germany.

Because no fossil intermediates lead from an unknown ancestor to pangolins, evolutionists use "just so" stories to fill in evolution's significant blanks. Ricki Lewis, a Ph.D. geneti-

cist, did exactly that with her article "How the Pangolin Got Its Scales—A Genetic Just-So Story." Lewis states:

> At some point in time, a few pangolins, thanks to chance mutations, had harder hair. Other mutations somehow guided those hairs to eventually overlap, providing shielding. Individuals whose hairiness began to become overlapping scaliness were less likely to succumb to bacterial infections, and thereby more likely to survive to pass on those traits.[2]

This kind of storytelling is common in evolutionary literature. Typical evolutionary descriptions are rife with imprecise, or even magical, terms like "chance mutations," "somehow guided," "began to become," "likely," "eventually," and "perhaps." Scales in mammals are clearly a novelty and require a unique explanation.[3] But since nature has no ability to select or "guide" anything, evolutinary explanations fail spectacularly.

Lewis also suggests in her article that pangolin armor through evolutionary time replaced part of the animal's immune system. She reasons the tightly knit scales protect the animal from predators and are also a barrier to bacteria. If so, why haven't all mammals in the same ecosystem evolved scales?

Pangolins are unique mammals engineered to move into and fill ecosystems in Africa and Asia. Fossils confirm the concept that Christ the Creator gave pangolins all their unique and fully integrated features right from the start. They look the same today as when they walked out of the Ark. And their mega-tongues and body scales reveal the same aspects of the Creator's ingenuity today as they did for Noah.

References

1. Benton, M. 2014. *Vertebrate Paleontology*, 4th ed. Malden, MA: Wiley Blackwell, 383.

2. Lewis, R. How the Pangolin Got Its Scales—A Genetic Just-So Story. *PLoS Blogs*. Posted on plos.org October 20, 2016, accessed November 5, 2016.

3. Arthur, W. 2011. Chapter 20 in *Evolution: A Developmental Approach*. Chichester, UK: Wiley-Blackwell.

3

WHEN FROGS, SNAKES, AND LIZARDS FLY

Some mammals, reptiles, and even amphibians can actually glide through the atmosphere. Christ's inventive engineering has equipped these unexpected animals for aerial travel. The fantastic designs of more familiar flyers like falcons and fruit bats shouldn't fail to inspire, but each newfound aeronautical wonder in the living world offers a fresh example of Christ's creativity.

Consider the so-called flying frogs. In Malaysia, the golden tree frog knows how to spread its arms and legs to control its descent from high in a jungle tree. Southeast of Malaysia, the Java flying frog uses webbed feet to resist air, slowing its descent even more. Indonesian jungles also host Wallace's flying frog, *Rhacophorus nigropalmatus*. Huge webbing between its toes and its aerodynamically flattened body allow it to glide at about a 45° angle.

A downside to having longer toes and extra webbing is that these features don't help with crawling or hopping. Therefore, each of these frogs strikes a unique balance between carrying the extra flesh needed to slow an airborne descent and having more nimble limbs to increase creeping agility. Thus, the Lord deserves praise for inventing the general concept of gliding frogs and credit for crafting different gliding grades that enable various tree frogs to fit and fill diverse jungle niches.

Not only can frogs sail, but certain snakes from parts of India can expertly glide through the air. Jake Socha, a flying-snake expert at Virginia Tech's Department of Engineering Science and Mechanics, summarized the major results from his experiments in a TEDx video.[1] He found that when the flying snake *Chrysopelea paradisi* travels through the air, it writhes first to one side and then the other so that its average body position is symmetrical when gliding in a straight line. It can also control tight turns by whipping its body around in midair. Without these skills, the animal would tilt sideways and tumble down.

Like a wing with adjustable flaps, aeronautical engineers would engineer a gliding snake to rotate and flatten its many ribs, "turning its entire body into a wing." Socha said, "This snake shape is able to generate a similar amount of lift to an engineered aerofoil. Not bad for a snake."[1] Of course, snakes don't engineer their own features any more than airplanes do. Our brilliant Creator Jesus Christ, not the snake, deserves all the credit.

When it glides through the air, the flying gecko *Ptychozoon kuhli* extends thin skin fringes that wrap around the lizard's sides. Plus, its skin comes camouflaged to mimic tree bark. With its standard gecko toe pads' microscopic fibers coated with superhydrophobic (water-repelling) lipids that "glue" them to almost any surface, these lizards pack plenty of purposeful design into a tiny package.[2]

Borneo's flying lizard *Draco cornutus* glides the farthest of all these creatures. It extends unique ribs that suspend skin webbing like what one would expect from a human-engineered retractable hang glider. It lives its whole life in Indonesian treetops, shifts its skin color from brown to green in active camouflage, and eats ants. If it lived in Peru, it might even eat gliding ants. Select species of tropical ants like *Cephalotes atratus* forage among treetops and can opt for a shortcut back to the trunk below by just jumping into the air![3] They recognize their tree trunk target, aim for it, and land expertly.

The Lord Jesus gets the credit for carefully crafting each of these gliding creatures because "by Him all things were created that are in heaven and that are on earth." We honor Him because "all things were created through Him and for Him."[4]

References

1. Socha, J. Snakes that fly—really. TEDx Virginia Tech. Posted on youtube.com December 6, 2012, accessed August 31, 2017.

2. Hsu, P. Y. et al. 2012. Direct evidence of phospholipids in gecko footprints and spatula–substrate contact interface detected using surface-sensitive spectroscopy. *Journal of the Royal Society Interface*. 9 (69): 657-664.

3. Yanoviak, S. P., R. Dudley, and M. Kaspari. 2005. Directed aerial descent in canopy ants. *Nature*. 433 (7026): 624-626.

4. Colossians 1:16.

4

BEES ARE REALLY, REALLY SMART

In 2005, biologists were stunned to discover that humans might not all look the same to honeybees. A study found that bees can learn to recognize human faces in photos and remember them for at least two days.[1]

Twelve years later, *Science* magazine published an article describing the "unprecedented cognitive flexibility" of bees using a ball to get a reward.[2] Biologist Olli Loukola said, "I think the most important result in our case was that bumblebees can not just copy others but they can improve upon what they are learning....This is of course amazing for small-brained insects—even for us, it's difficult to improve on something when we are copying others."[3]

It seems these tiny invertebrates have been engineered by Christ with the limited but extraordinary ability to reason and learn. In 2019, a study found that bees can link symbols to numbers. "We know bees get the concept of zero and can do basic math. Now researchers have discovered they may also be capable of connecting symbols to numbers."[4]

The researchers described their stunning bee results in *Proceedings of the Royal Society B*, saying, "Here we show that honeybees are able to learn to match a sign to a numerosity, or a numerosity to a sign, and subsequently transfer this knowledge to novel numerosity stimuli changed in colour properties, shape and configuration."[5]

And the honeybee accomplishes all this with a brain about the size of a sesame seed. A neuron is a nerve cell through which electrochemical impulses are transmitted. A whale's brain has over 200 billion neurons, human brains have an estimated 86 billion neurons, and a honeybee's (*Apis mellifera*) brain contains fewer than one million neurons. "Despite their tiny brains bees are capable of extraordinary feats of behavior," said Dr. Nigel Raine, from Royal Holloway's school of biological sciences at the University of London.[6] How right to think of such feats as extraordinary for natural processes to generate. But the Lord Jesus as a personal and ingenious Creator does extraordinary feats without effort.

Researchers are finding that it's not necessarily the mass or sheer number of neurons in an animal that accounts for the creature's cognition—it seems rather to lie in the neural circuits, specifically the circuits' interconnectivity and modularity.[7] It appears the neural circuits must be arranged and connected in a very specific manner in order to function as optimally as they do.

We see crude analogs in the way humans engineer data networks that link computers around the world. Neither God-made nor man-made networks would work without each module and switch being engineered to integrate for a specific task. Bee brains show expertly miniaturized neural networks that enable them to adapt on the fly—a necessary ability in a world with ever-changing pollen sources.

Bees are engineered with a tiny brain with less than one million neurons, and yet they can recognize human faces, count, improve upon what they are learning, do basic math, link symbols to numbers, and navigate using spatial memory with a "rich, map-like organization."[8]

As an evolutionist said, "These are, high, high, highly intelligent creatures."[3] One must ask: Are bees the result of time and chance or plan, purpose, and creation? The answer seems evident. With the Bible by our side, it's just as plain to identify Jesus as the one who organized bees' brains as it is to identify humans as the ones who organized computer networks.

References

1. Lucentini, J. Bees can recognize human faces, study finds. Physorg and World Science. Posted on phys.org December 11, 2005, accessed June 18, 2019.

2. Loukola, O. et al. 2017. Bumblebees show cognitive flexibility by improving on an observed complex behavior. *Science*. 355 (6327): 833-836.

3. Hugo, K. Intelligence test shows bees can learn to solve tasks from other bees. PBS News Hour. Posted on pbs.org February 23, 2017, accessed June 19, 2019.

4. Bees can link symbols to numbers, study finds. *ScienceDaily*. Posted on sciencedaily.com June 5, 2019, accessed June 19, 2019.

5. Howard, S. et al. 2019. Symbolic representation of numerosity by honeybees (*Apis mellifera*): matching characters to small quantities. *Proceedings of the Royal Society B*. 286: 1904.

6. Bees' tiny brains beat computers, study finds. *The Guardian*. Posted on theguardian.com October 24, 2010, accessed June 18, 2019.

7. Chittka, L. and J. Niven. 2009. Are Bigger Brains Better? *Current Biology*. 19 (21): R995-R1008.

8. Menzel, et al. 2005. Honey bees navigate according to a map-like spatial memory. *Proceedings of the National Academy of Sciences*. 102 (8): 3040-3045.

5

ENGINEERED PEPPERED MOTH COLOR CHANGES

One of the most persuasive evidences cited in textbooks in support of evolution is how the increasing coal soot during Britain's 18th- and 19th-century Industrial Revolution drove the color change observed in black peppered moths. But genetic findings raise questions about the accuracy of this evolutionary scenario.

The story is that only the white form of the peppered moth *Biston betularia* was known in Great Britain before widespread coal burning covered buildings and trees with soot. White-colored moths sunning themselves on blackened trunks were easy pickings for birds. Then a random mutation turned some moths black, and nature selected *against* white moths and *for* black moths, which dominated the population until the heavy coal-burning period ended.

But research indicates there may be genetic reasons to question whether the "mutation" causing black coloration was simply a lucky accident or something else. In 2016, a genetic research team led by Ilik J. Saccheri discovered that the black coloration was due to the insertion of a "transposable element" of DNA.[1] Transposable elements are mobile stretches of DNA that can be excised from one portion of a chromosome and inserted into a different location. These insertions can change the expression of genes.

Saccheri found that a large stretch of DNA exceeding 21,000 base pairs in length was inserted into another section of DNA that regulated the coloration of peppered moths. Of the 105 black moths the team examined, 103 (98%) had this identical insertion of the transposable element, but the insertion was absent in all 283 white moths studied. The repetitive insertion of the transposable element in the same location, along with its strong bias toward black moths, indicates this might be a regulated event and not a fortuitous accident. A later paper by Saccheri's team examined two other species that had a similar coloration response.[2]

Another observation challenges the idea that the peppered moth response to pollution was due to an accidental mutation. Saccheri said in a University of Liverpool news release:

> Although many people have heard about industrial melanism in the British peppered moth, it is not widely appreciated that dark forms increased in over 100 other species of moths during the period of industrial pollution. This raises the question of whether they relied on the same or similar genetic mechanism to achieve this colour change.[3]

These studies not only challenge the story that the black coloration was the result of an accidental mutation, they instead point to a regulated mechanism as the cause. Among their findings are: genetic hotspots for wing coloration, coloration master switches, transposable elements consistently located to the same gene regulatory region, and the rapid, widespread response across dozens of species.

As the evidence stands, the rapid and predictable responses by so many species of moths could be seen as support for an engineered response, one put in place by the Creator Jesus Christ.

References

1. van't Hof, A. E. et al. 2016 The industrial melanism mutation in British peppered moths is a transposable element. *Nature*. 534: 102-105.

2. van't Hof, A. E. et al. 2019. Genetic convergence of industrial melanism in three geometrid moths. *Biology Letters*. 15: 20190582.

3. 'Industrial melanism' linked to same gene in three moth species. University of Liverpool news release. Posted on liverpool.ac.uk October 16, 2019, accessed January 23, 2020.

6

BEETLE MOUTH GEARS SHOUT DESIGN

Beetles (order Coleoptera) are a common group of insects easily recognized by the pair of shiny forewings covering their body. Beetles make up almost 40% of described insects. If zoologists investigated just the Coleoptera, they would easily be busy well into the next century.

Japanese biologists discovered an astonishing structure within the mandibles (mouth pincers) of the horned beetle. The insects have complex gear-like structures that operate in "completely synchronous movements."[1] Entomologists at the Tokyo University of Agriculture and Technology reported, "A closer examination revealed that each mandible has two gear teeth, and the two sets mesh. As a result, when one mandible moves, so does the other."[2]

In 2013, *New Scientist* magazine reported unexpected machinery in insect larvae:

> The insect *Issus coleoptratus* is another animal with an unexpected bit of machinery hidden in its body. Its larvae are the first animals known to have interlocking gears, just like in the gearbox of a car.[3]

Evolutionists would explain that these complex movements and gears were not designed but slowly developed piecemeal. However, an evolutionary publication stated, "It might be that gears are easily broken, and as soon as one tooth is sheared off, the mechanism doesn't work as well."[3] If that's true, then how could a gearset evolve from wheels that didn't have their teeth?

Meshing gears demonstrate functional coherence—gears must align in order to function. A lone gearwheel serves no purpose. No real beetle would waste its energy building useless almost-gears. Humans purposefully engineer multiple gears to fit and work together from the top down, with the end in mind from the beginning. Same with beetle gears.

The synchronous movements of the mouth gears of the horned beetle shout design, plan, and purpose. What better attributes to use in describing God's only Son, "through whom also He made the worlds."[4]

References

1. Ichiishi, W. et al. 2019. Completely engaged three-dimensional mandibular gear-like structures in the adult horned beetles: reconsideration of bark-carving behaviors (Coleoptera, Scarabaeidae, Dynastinae). *ZooKeys.* 813: 89-110.

2. Marshall, M. Rhinoceros beetles have weird mouth gears that help them chew. *New Scientist.* Posted on newscientist.com February 1, 2019, accessed February 12, 2019.

3. Marshall, M. Zoologger: Transformer insect has gears in its legs. *New Scientist.* Posted on newscientist.com September 12, 2013, accessed February 12, 2019.

4. Hebrews 1:2.

7

AMAZING BLACK WIDOW WEB SILK

The sudden origin of spiders demonstrated in the fossil record and their amazing design both showcase biblical creation.[1,2] Spider fossils are rare. Only about 1,000 fossil species have been described worldwide, but they always display spiders' iconic eight legs, remarkably complex eyesight organs, and are, as creationists predict, 100% spiders. The Jurassic period was supposedly 165 million years ago, and yet spider fossils found in Northern China show spiders have changed very little since then.[3] This is because spiders didn't evolve from an unknown non-spider ancestor—they evidently didn't evolve at all.

Although invertebrate zoologists have known the structure of spiderweb fibers and the main sequence of amino acids that make up some spider-silk proteins, recent research has uncovered how black widow spiders (*Latrodectus*) produce their steel-strength silk webs. It's not a simple process by any means:

> Utilizing state-of-the-art techniques, the research team was able to more closely see inside the protein gland where the silk fibers originate, revealing a more complex, hierarchical protein assembly.[4]

The biologists went down to the nanoscale when they looked in the spider's silk glands to determine "the storage, transformation and transportation process involved in proteins becoming fibers." The key is in a supramolecular assembly of

very tiny units called *micelles*. Even when starting out, micelles are both complex and compound, which was not predicted. Researchers initially thought the black widow spider's remarkable fibers originated from a random solution. Instead, they found "hierarchical nano-assemblies (200 to 500 nanometers in diameter) of proteins stored in the spider's abdomen."[4]

The spider's web is a strand of compounded material woven and bonded together on a molecular scale the way human engineers would weave something like steel cable together for added strength. Researchers want to duplicate the spider's unique production of steel-strength fibers to make "equally strong synthetic materials" for use in "high-performance textiles for military, first responders and athletes; building materials for cable bridges and other construction; environmentally friendly replacements for plastics; and biomedical applications."[4] But they may never achieve the nanotechnological expertise the Lord Jesus gave these tiny spiders in the beginning.

Creationists give the Creator Jesus Christ—not time and chance—credit and praise for engineering spiders and their incredible silk production.

References

1. Sherwin, F. Spiders Have Always Been Spiders. *Creation Science Update.* Posted on ICR.org March 19, 2015, accessed November 9, 2018.

2. Thomas, B. Scientists Decode Key to Spider Web Strength. *Creation Science Update.* Posted on ICR.org March 19, 2012, accessed December 5, 2018; Thomas, B. The Masterful Design of Spider Webs. *Creation Science Update.* Posted on ICR.org March 30, 2012, accessed December 5, 2018; Sherwin, F. Spiral Wonder of the Spider Web. *Creation Science Update.* Posted on ICR.org May 1, 2006, accessed December 5, 2018.

3. Ghose, T. Stunningly Preserved 165-Million-Year-Old Spider Fossil Found. *Wired.* Posted on wired.com February 9, 2010, accessed November 15, 2018.

4. Mystery of how black widow spiders create steel-strength silk webs further unravelled. *ScienceDaily.* Posted on sciencedaily.com October 22, 2018, accessed November 15, 2018.

8

CAMOUFLAGE OCTOPUS SKIN

A mimic octopus can change the color and texture of its skin at will to match any kind of surrounding. It actively camouflages itself with astoundingly complicated biological machinery. Wouldn't it be great if, say, a soldier's uniform or an armored vehicle used similar camouflage technology?

Researchers from the University of Houston, the University of Illinois at Urbana-Champaign, and Northwestern University developed a heat-sensitive sheet that can automatically transition between black and white and produce multiple shades of gray. Achieving their tiny prototype was no easy feat. A University of Houston news report said:

> The flexible skin of the device is comprised of ultrathin layers, combining semiconductor actuators, switching components and light sensors with inorganic reflectors and organic color-changing materials in such a way to allow autonomous matching to background coloration.[1]

The team published their work in *Proceedings of the National Academy of Sciences*, where they described their design as inspired by octopus skin.[2] Just as the octopus must first detect the color of its surroundings, the photodiodes and multiplexing switches in this human-engineered material first detect background patterns. In a 44-page supplement to the

PNAS article, the authors listed 76 distinct steps required to fabricate their photodetector array and 74 steps to fabricate the diode array. Additional steps combine these arrays into the final product. But all this ingenuity pales in comparison to living octopus technology.

An octopus' biological photodetector is a sensor that continually tracks the environment in real time and seamlessly integrates with its near-infinitely variable shape and color outputs. A *Newsy* online video quoted one of the *PNAS* study authors as telling *National Geographic,* "As an engineer looking at movies of squid, octopuses, and cuttlefish, you just [realize] that you're not going to get close to that level of sophistication."[3]

If a single fabrication step fails to occur, then the man-made color-changing fabric would not function, illustrating the focused intent required to bring the design from conception to seamlessly functioning reality. How much more does the construction of octopus skin with its superior, higher-resolution, full-color fabric—that even heals itself—illustrate the focus and intent of a sophisticated ingenious Maker, our Lord Jesus Christ?

References

1. Kever, J. Researchers Draw Inspiration For Camouflage System From Marine Life. *University of Houston News*. Posted on uh.edu August 20, 2014.

2. Yu, C. et al. 2014. Adaptive optoelectronic camouflage systems with designs inspired by cephalopod skins. *Proceedings of the National Academy of Sciences*. 111 (36): 12998-13003.

3. Koeneman, B. Awesome New Camouflage Sheet Was Inspired By Octopus Skin. *Newsy*. Posted on newsy.com August 19, 2014, accessed August 19, 2014.

9

HORSESHOE CRABS: LIVING FOSSILS OR LIVING TESTAMENTS?

The horseshoe crab is a marine arthropod that testifies to the creative design of Christ's living creation. These odd crabs look a lot like crustaceans (lobsters, crabs, and shrimp) but belong to a group of the subphylum Chelicerata. They have a hard carapace (upper shell) and numerous eyes, including a pair of lateral eyes and median eyes that can detect ultraviolet light. Each eye has 1,000 ommatidia, the basic but highly organized unit of the arthropod compound eye.

Fossils of these creatures are found in the lowest layer of the geologic column—the Cambrian—where they and other sophisticated animal groups suddenly appear. For this reason, evolutionists call the horseshoe crab a *living fossil.* Recently, fossils of 100% horseshoe crabs were found in even earlier sediments.

> A batch of horseshoe crab fossils show [*sic*] that the critters are about 25 million years older [according to evolutionary dating] than was previously thought. What's more, the horseshoe crab fossils are incredibly complex, suggesting their ancestors evolved far earlier.[1]

During these alleged millions of years there has been no change in these animals, while subhuman creatures somehow became human, dinosaurs supposedly evolved into birds, and

a group of mammals even returned to the ocean to become whales. Isn't it strange that according to evolutionists, random mutations that are selected through deadly struggles to survive drove all those changes and yet didn't affect the horseshoe crab for all this supposed deep evolutionary time?

That's to be expected if nature doesn't really select, random mutations don't create novel genes from scratch, and those millions of years never happened. Creation scientists maintain horseshoe crabs have always been horseshoe crabs since their creation—that's why the fossils look fully formed right from the start. A world that's thousands not billions of years old leaves no time for crab evolution. No wonder horseshoe crab fossils match the living creatures.

Medical science discovered that the horseshoe crab's uniquely engineered blue opaque blood is very sensitive to contamination such as bacterial toxins.[2] The blood is used to test for impurities of various things that go into the human body, from implanted medical devices to injections and IV drips. Technologists are able to bleed horseshoe crabs and return most of them to their environment—shallow ocean waters or wet, sandy areas.

Biologists put together a biochemical test based on horse-shoe crab blood called *limulus amebocyte lysate (*LAL) that has replaced rabbits in testing for bacterial toxins. LAL is quick and effective but very expensive. A researcher named Jeak Ling Ding decided to produce an alternative to LAL that would take living horseshoe crabs off the hook. Ding wanted to utilize recombinant DNA technology—i.e., extracting DNA from one species and inserting it into another. This would involve splicing the LAL gene from the genome of the horseshoe crab and implanting it into rapidly growing, easy-to-maintain organisms such as yeast.

Scientists had identified the specific bacterial toxin-detecting molecule in LAL called *factor C*. After years of investigation and dead ends, Ding and a co-worker turned to insect-gut cells that could be used to easily produce low-cost factor C. Using a process called the *baculovirus vector system*, factor C was introduced into insect-gut cells that are, in effect, tiny factories for the factor C molecule. Evolutionists maintain that this is successful because both horseshoe crabs and insects are arthropods and supposedly share an evolutionary lineage.

Creationists counter that the phylum Arthropoda (the largest animal group, the evolutionary origin of which is unknown) was created and these animals have a common Designer, not a common ancestor. The fossil record bears this out. Horseshoe crabs aren't living fossils, they are living testaments to Christ's design. Meanwhile, the fact that humans can cleverly manage biological resources to help other people is a living testament to our having been created in God's image.

References

1. Geggel, L. Rare Fossils of 400-Million-Year-Old Sea Creatures Uncovered. *LiveScience.* Posted on livescience.com July 8, 2015, accessed May 10, 2018.

2. Zhang, S. The Last Days of the Blue-Blood Harvest. *The Atlantic.* Posted on theatlantic.com May 9, 2018, accessed May 10, 2018.

3. Geggel, L. 500-Million-Year-Old Brains of 'Sea Monsters' Get Close Look. *LiveScience.* Posted on livescience.com May 7, 2015, accessed May 10, 2018.

10

POLAR BEARS, FITTED TO FILL AND FLOURISH

Polar bears are furry giants fitted to fill frigid habitats in Arctic Ocean waters, ice floes, and shore lands. They provide four living lessons in apologetics—the defense of the Scripture that Paul spoke of, saying, "casting down arguments and every high thing that exalts itself against the knowledge of God, bringing every thought into captivity to the obedience of Christ."[1]

First, polar bear lifestyles corroborate biblical information. Scripture portrays bear behavior that matches what we observe today, including polar bear behavior. Female bears, as described in Scripture, are serious threats to anyone who angers them, especially someone threatening their cubs (2 Samuel 17:8; 2 Kings 2:24; Proverbs 17:12; Hosea 13:8). It's a risky adventure to fight a mama bear (1 Samuel 17:34-37)! Bears are omnivorous predators.[2] They growl (Isaiah 59:11), lie in wait for edible prey (Lamentations 3:10), and showcase fierceness (Daniel 7:5; Revelation 13:2). Hungry bears should be avoided until the time when God transforms them into strict vegetarians (Isaiah 11:7).

Second, polar bears' lives refute evolutionary speculations. Consistent with how Genesis reports biodiversity, creationists recognize an ursine "bear kind." Unsurprisingly, polar bears

can mate with other bears (e.g., polar bears breeding with grizzly/brown bears), yet this reality disproves earlier evolutionary notions of ursine speciation and genetic incompatibility.[3] Meanwhile, imaginary phylogenetic lineages—of bears with non-bears such as canines—are still missing the predicted transitional forms despite 150-plus years of extensive searching for them in the fossil record.[4]

Third, polar bears help clarify historical truth about global climate change. They aren't going extinct even if Earth warms up a few degrees, notwithstanding alarmist pseudoscience. They can safely survive vacillations of global climate change without any help from politicians. During the Medieval Warm Period lasting from about 950 to 1250 AD, polar bears (also called white bears or snow bears) survived. Vikings captured and marketed them as exotic animals.[5] After those "global warming" centuries, the cooler Little Ice Age followed from approximately 1350 to 1850. Polar bears survived again.

Fourth, polar bears glorify Jesus as their Creator simply by living their lives. Like other wild beasts in God's created world (Revelation 4:11), polar bears daily demonstrate God's caring providence just by being themselves. For example, although baby polar bears are conceived during the spring, uterine implantation of embryos—as with other bears as well as mustelids (e.g., weasels) and seals—is delayed by design until autumn. That's when mama bear enters her maternity-ward den, ensuring that births occur in winter during hibernation. The family's den exodus is timed for spring, when food availability is optimal and infant cubs are physically developed enough to travel on sea ice.[2]

Finally, consider the energizing nutrition God installed in polar bear mothers. Polar bear babies are born small, about 1.5 pounds—one-fifth the size of human babies. Before leaving the den in spring, each cub needs to weigh around 25 to 30 pounds. Following the initial protein-loaded, antibody-rich colostrum, milk for newborns can be 46% fat, facilitating a get-big-and-fat-quick growth pattern. Yet, fat content declines over time to about 5% (like in human milk) at weaning, having fueled a 1,500 to 2,000% weight gain during three to four months.[3] God's design delivers precisely what's needed.

In every physiological area, polar bears are optimally suited to their environment. What's more, Christ engineered the original bear kind with the potential to diversify and fill Earth's ever-changing environments and climates. From the bears on Noah's Ark came polar, black, cave, and other bears. God's ingenious plan for provision delivers just what our changing world demands. What a worthy Creator we have in the Person of Jesus Christ!

References

1. 2 Corinthians 10:5. These four apologetics priorities—supporting Scripture, eroding evolution, clarifying climate confusion, and glorifying God—are priorities in the ICR Discovery Center for Science & Earth History's exhibits in Dallas, Texas.

2. Cansdale, G. S. 1976. *All the Animals of the Bible Lands.* Grand Rapids, MI: Zondervan, 17-18, 27, 109-110, 116-119; Derocher, A. E. 2012. *Polar Bears: A Complete Guide to Their Biology and Behavior.* Baltimore, MD: Johns Hopkins University, 155, 172-180. Creation evidences defend the faith by corroborating Bible-reported information.

3. Roach, J. Grizzly-Polar Bear Hybrid Found—But What Does It Mean? *National Geographic News.* Posted on nationalgeographic.com May 16, 2006, accessed June 5, 2017. Creation evidences routinely impeach and refute evolutionist errors such as materialism and animism in natural selection mythology.

4. Morris, J. D. 2006. What's a Missing Link? *Acts & Facts.* 35 (4).

5. Logan, F. D. 2005. *The Vikings in History*, 3rd ed. London: Routledge, 58-61. Correcting confusions caused by uniformitarianism is yet another priority for biblical creation apologetics.

11
HUMMINGBIRDS: TINY AEROBATIC EXPERTS

Who doesn't pause to marvel when a hummingbird flies by? These tiny, colorful birds perform amazing aerobatic feats, and yet some very smart scientists insist that mere natural forces mimicked a real engineer to construct these fascinating flyers. Authors of a *Nature* paper on hummingbird flight wrote in 2005 that "the selective pressure on hummingbird ancestors was probably for increased efficiency."[1] They imagine that hummingbirds evolved from ancestors that could hover only briefly.

But just as nature has no agency to select anything, a creature's environment has no foresight, nor can it produce a real force to selectively "pressure" the creature toward "increased efficiency" or any other trait. An examination of just a few key hummingbird features leaves no doubt "that the hand of the LORD has done this."[2]

Hummingbird beaks, bones, and feathers differ from those of all other living or extinct bird kinds.[3] Their wings don't fold in the middle. Instead, they have a unique swivel joint where the wing attaches to the body so that the wings rotate in a figure-eight pattern. And they move fast! They have to beat their wings rapidly to hover, levitating with level heads as they extract nectar from flowers for hours per day. Scientists still need to discover the bird's mental software that coordinates information about the location of a flower's center with muscle motion that expertly stabilizes the hummingbird's little head as it drinks.[4]

Its long, slender beak and skinny tongue dip into and out of the flower to gather nectar using a clever automatic fluid-trapping mechanism. Tiny, curved structures along the tongue's tip open to hold nectar, then curl up tightly after the bird swallows.[5] When the hummingbird finishes with one flower—or with the backyard hummingbird feeder—it moves away by flying backwards. It could not do this, nor could it twist, dive, or maneuver through the air the way it does, without having extra-long primary feathers on its wings. These are its largest body feathers and produce most of the needed lift.

Could chance natural processes transform a bird like a treeswift into a hummingbird by adding required parts one at a time? Imagine a bird that had a hinge joint, long primary feathers, and head-balancing and body-leveling circuitry, but it still had a short beak with a short tongue to fit—or even a long tongue that didn't fit. Such a creature might hover in front of a flower but could never reach its food without a suitable beak. Wouldn't such a partly evolved creature starve to death before imaginary selective "forces" could add the right beak?

Even if a hummingbird somehow acquired every flight-required part except one—say, its primary feathers were a centimeter too short, or it had everything in place except its unique hinge joint—the creature could not fly. The hummingbird reflects all-or-nothing unity in its physiological attributes. This functional coherence must have been engineered and built into the very first hummingbird or it couldn't fly and reproduce.

Ongoing hummingbird research has revealed other fascinating features. Birds generate a lot of heat when they fly. Con-

sidering their speed, you might expect hummingbirds to burst into flames at any moment. Where does all that body heat go? Infrared cameras revealed hummingbird "radiators" that direct body heat out through the feet, shoulders, and eye areas.[6] And some male hummingbirds use air flowing through their tail feathers to produce melodious sounds during courtship.[7]

Our great Creator the Lord Jesus Christ ingeniously integrated all these phenomenal features into His tiny aerobatic experts.

References

1. Warrick, D. R., B. W. Tobalske, and D. R. Powers. 2005. Aerodynamics of the hovering hummingbird. *Nature.* 435 (7045): 1094-1097.

2. Job 12:9.

3. Hummingbird fossils look like modern hummingbirds. See Mayr, G. 2004. Old World Fossil Record of Modern-Type Hummingbirds. *Science.* 304 (5672): 861-864.

4. They also eat spiders and insects for proteins and lipids.

5. Rico-Guevara, A. and M. A. Rubega. 2011. The hummingbird tongue is a fluid trap, not a capillary tube. *Proceedings of the National Academy of Sciences.* 108 (23): 9356-9360. Also see the online video FLIGHT: The Genius of Birds-Hummingbird tongue. Illustra Media. Posted on youtube.com June 18, 2013, accessed January 26, 2016.

6. Powers, D. R. et al. 2015. Heat dissipation during hovering and forward flight in hummingbirds. *Royal Society Open Science.* 2: 150598.

7. Clark, C. J., D. O. Elias, and R. O. Prum. 2011. Aeroelastic Flutter Produces Hummingbird Feather Songs. *Science.* 333 (6048): 1430-1433.

12

BAT BIOSONAR SEES IN THE DARK

Picture a calm summer evening. Most people are only dimly aware of the aerial creatures that might dart and dive nearby. Indeed, many would think these animals are birds unless their activity is closely observed.

They'd be surprised to discover these flying creatures are actually bats using their engineered sonar to detect and track insects as small as mosquitoes—prey these bats pick up and devour on the wing and in the dark. Looking at the amazing design features of a typical bat, Douglas Futuyma assumes "the only scientific explanation of adaptations is the theory of evolution by natural selection."[1]

But how could nature, which lacks both volition and agency, "select" anything? There is another, far more logical explanation of bat origins found in Genesis. Looking at the fossil record, we find that bats have always been bats. There are no alleged evolutionary intermediates between any four-legged rodent and today's two-legged, two-winged bat; there are no fossils of non-flying bats. Supposedly, the oldest fossil bats are from Wyoming (e.g., *Icaronycteris* and *Onychonycteris*) and Europe, dated by evolutionists to the early Eocene. Neither can evolutionists explain the exquisitely engineered bat biosonar that uses sound waves rather than radio waves. Michael Benton states, "The evolution of echolocation in bats has been hard to resolve."[2]

Hard to resolve or not, zoologists have discovered bats use this remarkable biosonar echolocation system to determine the texture, size, density, and shape of an object. "A bat builds a mental image of its surroundings from echo scanning that approaches the resolution of a visible image from eyes of [daytime] animals."[3]

Research has documented this astonishing claim. Scientists at the Universities of Munich and Bristol discovered a unique mechanism bats use to perceive object size.[4] Bats are able to detect tiny timing differences between when a sound reaches each ear. Sounds arrive from diverse directions, and bats use this mechanism to build a virtual image with remarkably detailed features. A press release reporting on these studies states that this gives the "bats access to comparable information about objects as we obtain with our eyes."[5]

Sensing reflected sound waves from a moving object in the dark, building a detailed mental image, pinpointing its location in relation to the hunting bat, and then catching the insect all while flying requires many intricate systems working together—sonar emitter, sonar detector, sonar data processor, logic outputs, and data networks. All these systems must have been engineered and put in place at the beginning.

One of the more fascinating relationships in the animal world is the ultrasonic interaction between moths and bats. The Creator engineered in some moths (e.g., the hawk moth) and a host of other insects the ability to pick up specific phases of bat sonar by way of unique cells imbedded in the insect's body. When sonar impulses strike the moth's body, these receptor cells send a message (action potential) up to the moth's brain that alerts it to bat activity nearby. The moth knows it's in danger and takes immediate evasive action.

Recently, it was discovered that when some types of bats

encounter these evasive moth maneuvers, the bats engage in countermeasures. Specifically, in the last phase of their pursuit, the bats widen their beam of echolocation to keep their supper within "sight." Jakobsen et al stated, "Thus, beam broadening is not a general property of echolocation, but we hypothesize that maintaining a broad acoustic field of view is crucial for all echolocators hunting moving prey."[6]

Jesus deserves credit for gifting bats with the ability to see in the dark. Future research into this amazing bat biosonar promises to further verify the "clearly seen" creative hand behind it (Romans 1:20).

References

1. Futuyma, D. J. 2013. *Evolution*, 3rd ed. Sunderland, MA: Sinauer Associates, Inc., 1.

2. Benton, M. J. 2005. *Vertebrate Palaeontology*, 3rd ed. Malden, MA: Blackwell Publishing, 376.

3. Hickman, C. et al. 2011. *Integrated Principles of Zoology*, 15th ed. New York: McGraw-Hill, 631.

4. Goerlitz, H. R., D. Genzel, and L. Wiegrebe. 2012. Bats' avoidance of real and virtual objects: implications for the sonar coding of object size. *Behavioural Processes*. 89 (1): 62-67; Heinrich, M. et al. 2011. The sonar aperture and its neural representation in bats. *Journal of Neuroscience*. 31 (43): 15618-15627.

5. How bats 'hear' objects in their path. University of Bristol press release. Posted on bristol. ac.uk November 24, 2011 accessed June 26, 2015.

6. Jakobsen, L., M. N. Olsen, and A. Surlykke. 2015. Dynamics of the echolocation beam during prey pursuit in aerial hawking bats. *Proceedings of the National Academy of Sciences*. 112 (26): 8118-8123.

13

THE DEXTEROUS DRAGONFLY

No flying machine or other creature has the aerial dexterity of the dragonfly. It can fly upside-down and backwards as easily as straight ahead. And it moves so fast that researchers have to use high-speed cameras to study it. A recent report asserted that dragonflies achieved their flying skills because they have had millions of years to perfect and hone them. But is this scientifically grounded or just a flight of fancy?

Harvard University biomechanist Stacey Combes and her team studied the way dragonfly flight operates. In a video posted by *Science Nation,* an online magazine funded by the National Science Foundation, one dragonfly with half of its right wing removed successfully caught a fruit fly in flight. No man-made aircraft of any kind can fly with the same kind of extensive wing damage. When it came to explaining dragonfly origins, the report said:

> Dragonflies have had a long time to evolve their skills as predators. They have been on the planet for about 300 hundred [*sic*] million years and predate dinosaurs. They can fly straight up, straight down, hover like helicopters and disappear in a blur.[1]

But does any scientific observation document how these kinds of predatory skills—or any skills or traits of any creature, for that matter—supposedly evolve? Does any experiment

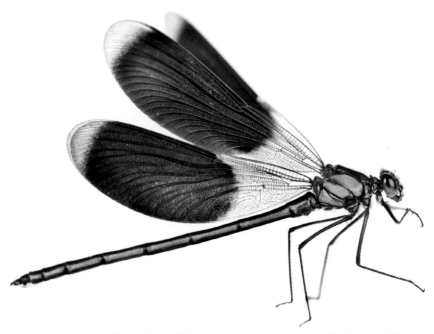

demonstrate that the addition of more time yields better flying (or other equally specific) structures in animals? And does
credible evidence substantiate the "300 million years" claim?

It's true that dragonfly fossils are found in sedimentary rock
layers below those containing dinosaurs. However, if most fossils were deposited in just one year by the worldwide Flood of
Noah, then the geologic layers don't represent different time
periods but different biomes.[2]

In other words, some swamp-dwelling dragonfly habitats
were inundated prior to the more terrestrial habitats that con-
tained the dinosaurs, conifers, birds, and mammals that are
found together as fossils. But since the rock layers show ca-
tastrophe and are broad in extent, both the layers containing
dragonfly fossils and those with dinosaurs appear to have been
deposited as phases within the overall Flood year.

In addition, researchers can test the idea that dragonflies

evolved their flying skills. Dragonflies were catastrophically buried in mud—which evidently flowed faster than the insects could fly—that later turned to stone. If the dragonfly fossils show partially evolved features, then that would support the claim that they might have evolved. But their fossils instead look identical in core design and construction to dragonflies living today. That, of course, supports the idea that they were created perfectly equipped for flight from the beginning.

Give a blind and dumb tinkerer millions of years in a scrap yard, and instead of getting closer to crafting an airplane, the bits of scraps will have gone with the wind. The idea that a long time helped dragonflies "evolve their skills" robs the Lord Jesus of the credit He deserves for crafting these expert flyers from the start and gives that credit to nature, which is idolized in evolutionary literature as a substitute creator.

The biggest difference between modern and fossil dragonflies is that many of the fossilized ones were several times larger, some having wingspans of over three feet! If anything, dragonflies have devolved into smaller versions, not evolved.

So, there's no fossil evidence whatsoever that dragonfly flight evolved. In fact, since fossilized and living dragonflies share the same structure, they show no sign of either millions of years or evolution. This is because like man-made flying machines, dragonflies were recently and purposely engineered by the Master Engineer revealed in the Bible as the Lord Jesus Christ.

References

1. O'Brien, M. and A. Kellan. Dragonflies: The Flying Aces of the Insect World. *Science Nation*. Posted on nsf.gov October 3, 2011, accessed October 13, 2011.

2. Parker, G. 2006. *The Fossil Book*. Green Forest, AR: Master Books, 21.

THIS BOOKLET WAS ADAPTED FROM THE FOLLOWING MATERIALS.

Thomas, B. Monarch Butterfly Antenna: A Hi-tech Tiny Toolkit. *Creation Science Update.* Posted on ICR.org October 9, 2009.

Sherwin, F. The Pangolin: A Mammal with Lizard Scales. *Creation Science Update.* Posted on ICR.org November 17, 2016.

Thomas, B. 2019. When Frogs Fly. *Acts & Facts.* 47 (1): 13.

Sherwin, F. Bee Brains Aren't Pea Brains. *Creation Science Update.* Posted on ICR.org July 11, 2019.

Guliuzza, R. J. Peppered Moth Color Changes Are Engineered. *Creation Science Update.* Posted on ICR.org February 25, 2020.

Sherwin, F. Beetle Mouth-Gears Shout Design. *Creation Science Update.* Posted on ICR.org March 12, 2019.

Sherwin, F. Amazing Design of Black Widow Web Silk. *Creation Science Update.* Posted on ICR.org December 6, 2018.

Thomas, B. Octopus Skin Inspires High-Tech Camouflage Fabric. *Creation Science Update.* Posted on ICR.org August 27, 2014.

Sherwin, F. Horseshoe Crabs: Living Fossils or Living Laboratories? *Creation Science Update.* Posted on ICR.org June 28, 2018.

Johnson, J. J. S. 2017. Polar Bears, Fitted to Fill and Flourish. *Acts & Facts.* 46 (8): 21.

Thomas, B. 2016. Hummingbirds! *Acts & Facts.* 45 (4): 16.

Sherwin, F. 2015. The Ultrasonic War Between Bats and Moths. *Acts & Facts.* 44 (10): 15.

Thomas, B. Did Dragonflies Really Predate Dinosaurs? *Creation Science Update.* Posted on ICR.org October 20, 2011.

IMAGE CREDITS

Bigstock Photo: cover, 1, 4, 8, 10, 14, 18, 23, 28, 31, 34, 42, 46

iStockphoto: 25, 39